Margaret and Robert Donington

SCALES, ARPEGGIOS, AND EXERCISES FOR THE RECORDER

(*SOPRANINO, DESCANT, TREBLE, TENOR, and BASS*)

OXFORD UNIVERSITY PRESS

MUSIC DEPARTMENT · WALTON STREET · OXFORD OX2 6DP

© *1961, Oxford University Press*

Photocopying this copyright material is **ILLEGAL**

Printed in Great Britain
Halstan & Co. Ltd., Amersham, Bucks., England

CONTENTS

		Page
1.	How to use this book	1
2.	Chart of fingerings	8
3.	Notes on clefs and pitch	9
4.	Notation of registers	10
	PART I	
5.	Scales and arpeggios for descant and tenor recorders	11
6.	Exercises for descant and tenor recorders, with notes on the special difficulties encountered in them	26
	PART II	
7.	Scales and arpeggios for sopranino, treble, and bass recorders	43
8.	Exercises for sopranino, treble, and bass recorders, with notes on the special difficulties encountered in them	58
9.	Index	75

HOW TO USE THIS BOOK

WHY THIS BOOK?

This is a book to help you as quickly and pleasantly as possible over the fundamental difficulties of playing the recorder.

WHY SCALES?

Because there is nothing else which serves quite as well if you want to get your fingers and your breath working reliably—so reliably that you can forget you ever had to take any trouble to get that reliability.

Scales and Arpeggios

Scales take care of the steps from note to note. Broken chords (arpeggios) have a use nearly as important. They take care of the leaps.

Special exercises

Since every instrument has its own awkward corners, and the recorder is no exception, there is a further possibility of saving precious practice time. This is by devising special exercises to take care of any steps or leaps which are particularly tricky.

Two books for further study

Beyond this point, there are many ingenious expedients for reducing technical difficulties by linking them with the kind of musical situation in which they are likely to arise. But these are not the fundamental difficulties, and they lie outside what we have to offer here. What we have to offer here is practice material for the foundation on which good playing can be developed in a more general way.

For that more general development, we recommend you to *Recorder Technique* and *A Practice Book for the Treble Recorder*, both by A. Rowland-Jones, O.U.P., London, 1959, 1961.

HOW TO USE THIS BOOK

Even technical practising can be enjoyed. We have taken great pains not only to fit our material to the recorder, but to give this material a shapely turn. It is all very simple; but even scales and exercises can be made into melodies with a musical rise and fall.

Stick at them, then, so long as your mind is well and truly on them and you are taking pleasure in the challenge. If they begin to irritate you, give them a rest, because no one ever did his technique any good by getting bored and practising his mistakes.

The material complete for those who want it

The convenience of scales is that they can be taken just as far as you require. We have provided good working scale patterns in all keys, major and minor, with broken chords (arpeggios) and special exercises to correspond. Those who want a diploma from the Trinity College of Music or similar bodies insisting on comprehensive professional standards for the recorder will need to work their way through the entire series. By the time they have done this satisfactorily, they should have the technical resources to meet any situation. Naturally their studies will have extended in a great many directions by then, and their musical experience will be wide and varied.

But selected parts can be used effectively

Other players may not be looking so far ahead. Almost all early recorder music, and a considerable proportion of modern recorder music, is in the simplest keys. Yet to play it really well, these keys must be mastered just as thoroughly as the remoter keys have to be mastered by a fully trained professional.

Those of you who only need to learn up to perhaps three flats or sharps can develop just the same solid reliability and easy style by working steadily through to that point and no farther. You are limiting your resources (as well as your problems) within a narrower range, but that need not prevent you from making a workmanlike approach.

SUGGESTIONS FOR PRACTISING
Adapting the order to your own problems

We have done our best to grade the difficulties of this book to lead on, by comfortable stages, from easier to harder. We have planned these stages as we might plan them in teaching an average pupil. But pupils differ very much in what they find difficult. It may be the top notes, or the bottom notes; it may be cross-fingering or half-holing.

We therefore recommend each player to adapt the order of study to suit his own requirements. We recommend teachers to bear the same point in mind.

Nevertheless, the order should still be systematic. It is not a good idea to turn haphazardly from one page to another. Least of all is it a good idea to work through the book in too much of a hurry. It saves time in the end if no difficulty is left behind until it has at least become a great deal less difficult than it was before.

Using the special exercises

At the top of each scale or broken chord you will find the numbers of certain special exercises. These are the exercises concentrating on the particularly awkward corners which you are going to have to negotiate. They are the hardest bits.

We suggest that you start by trying out the special exercises thus recommended. If they go easily, so will the scale or broken chord. More probably they will prove distinctly tricky. Give them your best attention for a time. Then turn back to the scale or broken chord itself.

When equivalent sounds are written differently

We are not trying to teach you to become a good reader. That is a different aspect of technique, and has to be approached by a different road. We are trying to teach you to use your fingers, your tongue and your breath with such easy familiarity that it becomes second nature.

If there is an exercise, for instance, which trains your fingers to work well between F and A flat, you will not necessarily find one to train them from E sharp to G sharp. That is all the same to your fingers, though not to your eye or your mind. On the other hand, since such enharmonic alternatives do look very different, and since if the eye or the mind is puzzled the fingers are bound to lag behind, we have in fact quite often included them.

We want you to notice when your fingers are making the same movements, and the recorder is making in effect the same sounds, although the written notes are different. That will take a lot of the sting out of the remoter keys. They may look black with flats or sharps, and yet the movements your fingers have to make are no different from those which they have already learnt. Once you have linked up the familiar movements with the unfamiliar keys, the battle is half won.

How to get round troublesome difficulties

We advise you never to let yourself get really stuck in a difficulty. Give it a reasonable amount of pertinacity and patience: it may work out sooner than you think, but if it will not work out, leave it alone. Go on to the next problem: the change will do you good. By all means run away and live to fight another day, but come back fresh and begin at the beginning again: feel your way into the notes, never rushing at them, never getting mechanical, but keeping your mind on a little bit at a time.

HOW TO BE CUNNING IN YOUR WAY OF PRACTISING

There is no cunning in charging straight through an exercise. The cunning lies in feeling your way all round it first. Try out a finger-movement or two at a time. Think the movement quite clearly beforehand; see it quite vividly in your mind's eye. Then let it happen almost of its own accord. It may not come at once, but presently it will, and then you can go on to the next bit.

Telling your muscles in good time what to do

The secret is to get on good terms with your muscles. By forcing them you make them stiff, so that they fight one against the other, and then you have a business persuading them to relax again. It is much easier to take them into your confidence from the start. Try not to bully your fingers. First think into them what they have to do, and next let them do it quietly for themselves.

It is the same with your tonguing. Tell your tongue in good time what you are asking it to do. Then it has no need to lag behind and be flabby. Neat, small movements are required. Think the rhythm into your tongue like a dance, and encourage it to do its job in a free and easy way.

Keeping in step

If you treat your fingers and your tongue considerately they are not so likely to get out of step, but they need some help in this respect, and there are ways of giving this help tactfully.

When it is a case of the fingers and the tongue being out of step with one another, it is quite often the tongue which causes the trouble by getting ahead. So leave your fingers to look after themselves for the moment, and give all your attention to your tongue.

In cross-fingerings (that is to say when there are fingers going down on to their holes and fingers coming up from their holes at the same time) if the fingers get out of step, it is almost invariably the up-coming fingers which cause the trouble, in this case by lagging behind. Leave the down-going fingers to look after themselves for the moment, and give all your attention to the up-coming fingers. These are the ones which seem to want to stick and get late.

In the right hand (though not in the left) you are using your fourth (little) finger, and this is not by nature so strong and manageable as the others. Encourage it by giving it extra practice. To a lesser extent this applies to the third finger (on either hand) as well.

Fingerings which require the third and fourth fingers of the right hand to move together also need extra practice. If they get out of step put your attention on the little finger. To a lesser extent the second and third fingers of either hand need similar treatment, with your attention on the third finger.

The thumb is not given a number in describing fingers for the recorder. The index finger is therefore described as the first; the next two fingers are described as the second and third; and the little finger is the fourth—i.e. on the right hand (the left hand does not use it).

FURTHER FINGERING HINTS

The bottom note of the recorder is difficult to get, particularly when the previous note leaves a lot of fingers in the air, and all these fingers have therefore to be brought down securely on to their holes at the same time. To overcome this difficulty try putting your fingers down one at a time, starting at the upper end of the recorder. Then try putting them down one hand at a time, starting with the upper hand. Then try putting them all down at the same time: quite gently, yet firmly too. If there is a "leak" somewhere, the bottom note will not sound. See if you can feel which finger is the culprit, by trying them separately again. Then give that culprit all your attention for the moment—by *feel* and not by *sight*. Use the sensitive pad of your fingers as a blind person uses it to read braille.

Half-holing for the bottom notes

The bottom semitones are also difficult. These are: C sharp (D flat) and D sharp (E flat) on the descant and tenor recorders; F sharp (G flat) and G sharp (A flat) on the sopranino and treble recorders. The bass recorder has a key for its bottom note, which makes the semitone above it impossible. This is also the case with some tenor recorders.

Except where there are keys, most of the best recorders have two little holes side by side for these notes, one of which (the farthest from your hand) has to be uncovered for the semitone in question. Where there is only one hole of the full size, half of it (the farthest half) has to be uncovered. That is a little harder to manage accurately; but the movement is the same.

Throughout this book "half-holing" means *either* half-covering a full-sized hole, *or* covering the nearer of a pair of little holes.

One way to make the movement for half-holing is to aim directly at the single little hole (or half the full-sized hole); but this is a tiny target which it is easy to miss. There is another method which you may find more secure. This involves slapping your finger fully down and then sliding or tipping it half-way back, so quickly that no one can detect it.

Never waste a movement of the fingers

Except for this special movement, which is only needed for these bottom semitones, the fingers go straight down on to their holes and stay there until they need to move for the next note. Do not make unnecessary movements of the fingers: they only render the necessary ones more difficult. When a finger is going to be needed again on the next note, keep it down.

Fingers which can be used as pivots

In two instances a finger can act as a pivot between the fingers on either side of it: this is a trick worth practising specially.

In the right hand, the second finger may have to stay down while the first and third fingers move alternately up and down (to get from E to F sharp or back on the descant and tenor; from A to B or back on the sopranino, treble and bass). Think of this second finger as the pivot.

In the left hand, the first finger has to stay down while the thumb and the second finger alternate (to get from B to C sharp or back on the descant and tenor; from E to F sharp or back on the sopranino, treble and bass). Here it is this first finger which you must think of as the pivot.

This pivoting is most easily done by a very slight rocking movement of the hand, keeping the wrist loose enough to turn freely.

How hard to press the fingers down

Whatever the fingering, press just hard enough to make a soft, air-tight covering, using not the sharp tip of the finger, but its nice broad pad. Any unnecessary pressure simply makes it harder to let the fingers up again.

Keeping the fingers ready for action

When the fingers come up, poise them delicately above their holes, ready for future action as required. They must be at least half an inch away so that they do not interfere with the free flow of air from the uncovered holes (which would send you flat); but they should not be much farther away (which would unnecessarily increase the distance through which they have to move).

How quickly to move the fingers

When the fingers do move, see that they move in a purposeful manner: they should not jerk; yet they should work positively. Give them a definite action as though they knew where they were going. Let them feel as if they were well oiled.

THE THUMBS

Even the thumbs can share in this well-oiled feeling.

Keeping the right-hand thumb relaxed

The right-hand thumb has nothing to do except help in supporting the recorder. But spare it a little thought from time to time. It is very important to keep it relaxed. If it gets stiff, this stiffness gets into the fingers too, and the whole hand has difficulty in working freely. When your fingers seem more obstinate than usual, it is a good idea to find out whether this obstinacy is starting in the right-hand thumb. Catch it unawares, and if you find it is rigid, that is where the trouble lies.

Keeping the left-hand thumb ready for action

This relaxed feeling is still more important with the left-hand thumb, which besides helping to support the instrument has to half-stop the thumb hole for the notes of the upper octave.

This half-stopping is done by putting the thumb-nail into the hole. The best way of achieving this is by bending the top joint of the thumb and unbending it again, with some slight assistance from the second joint. The need to keep these joints, especially the top joint, "well oiled" is therefore obvious. In other words, the muscles, though not allowed to go flabby, must not be tensed either. They must just be pleasantly relaxed, like all the other muscles used in playing the recorder.

TONGUING

Tonguing interrupts the breath between one note and the next. This separates the notes, and gives each a little impulse of its own; but the separation is usually very slight, and the impulse only enough to give them life and character.

This is the medium degree of articulation, and the most commonly useful. A much more distinct staccato (detached style) is also possible, and so is a legato (slurred) flow with no tonguing and therefore no separation at all.

Every possible variation in *tonguing* should be used in playing the recorder to compensate for the fact that very little variation in *loudness* can be made without getting out of tune. In this matter of articulation, the tongue is to the recorder player what the bow is to the violinist: it is what makes the player able to phrase his music in an interesting and expressive way.

Standard methods of tonguing

Standard tonguing may be done by putting your tongue behind your teeth, and immediately taking it away again, as in producing the letter T. For a softer effect, you can use the letter D instead. Paradoxical though it may seem, the more smoothly you wish to play, the more rapid the tongue movement must be (in order not to interrupt the flow of sound too much): rapid, yet at the same time gentle (in order to get the smoothness). But even for a sharper articulation, avoid a spitting action, which sounds much too violent whatever the nature of the music.

Methods of double-tonguing

Rapid notes can often best be tongued in pairs. To do this, you alternate two syllables: Te-Ke or De-Ge (Diddle for very soft notes). This is called "double-tonguing". Triple tonguing, saying Te-Ke-Te, or De-Ge-De, may also be used.

How to practise slurs

We suggest that you start each scale, broken chord or exercise by tonguing all the notes separately. Then practise slurring two notes in succession; or four notes; or more; or a mixture of tongued and slurred notes in various patterns. But always decide your pattern in advance, and stick to it until you decide to make a change.

Tonguing (i.e., separating the notes) shows up how well your tonguing is in step with your fingering. Slurring (i.e., joining the notes) shows up how clean the fingering itself is.

We have marked sample slurs in some of the scales and exercises. But we advise you also to experiment with varied slurrings of your own. Treat the printed slurs only as suggestions.

Always start with slow practice; but when your fingering can keep up with a rapid speed, it is time to introduce double tonguing as well as single.

THE BREATH

Of all the factors in good recorder playing the breath is the most vital, because it makes the actual sound.

If you are a singer, you will know about controlling and supporting your breath from the diaphragm. If not, it would be helpful to have some lessons from a good singing teacher.

How your breath can keep you in tune

Too little breath on the recorder sends you flat, and also makes a feeble sound. Too much breath sends you sharp, and also makes a piercing sound. Yet it is wonderful how much you can reduce the breath for a soft passage, provided you know how to keep the flow of air somehow comfortable and relaxed.

How to fit your breathing into the phrasing

Be sure to take a deep breath before starting to play. This gives you a good supply in reserve. Do the same whenever there is a rest or pause in the music long enough to offer you the opportunity.

But in the middle of the music, use every chance to take even a small breath in order to replenish the supply. It is better to take many small breaths in good time, rather than let the supply run low. Then there will be no need for an obvious gasp at the last moment.

Be careful to take even these small breaths at places that fit in with the musical phrasing, so that they do not disturb the natural flow of the music.

Try to go on pouring your breath down into the recorder in a round, smooth and steady stream as if it were so much treacle, relying on your tongue to point the phrasing. That should give you a round, smooth and steady tone.

It is this roundness of tone which can make the recorder one of the most beautiful of instruments.

CHART OF FINGERINGS FOR THE RECORDER

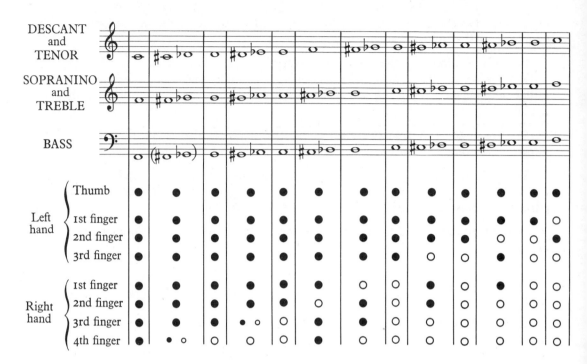

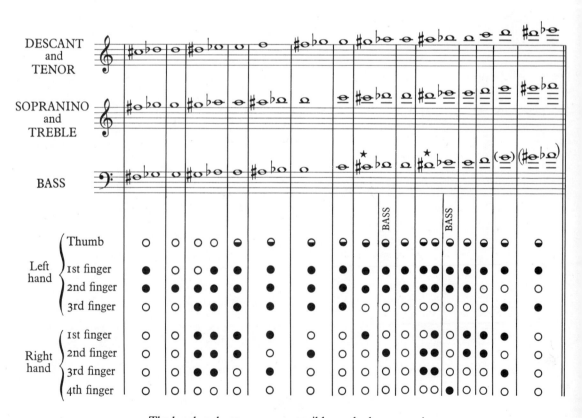

The bracketed notes are not possible on the bass recorder.
★ These notes require exceptional fingerings.
A few other high notes are not possible on some bass instruments.

NOTES ON CLEFS AND PITCH

Parts for the **treble** recorder are written in the treble clef. They sound at the pitch at which they are written. The music in Part II is designed for the treble recorder.

Parts for the **tenor** recorder are also written in the treble clef. They also sound at the pitch at which they are written. But the same fingerings produce notes a fourth lower than they produce on the treble recorder. The music in Part I is therefore designed for the tenor recorder. It is the same as the music in Part II, but written out a fourth lower.

Parts for the **sopranino** recorder are written in the treble clef, but sound an octave higher than the pitch at which they are written. The same fingerings produce notes an octave higher than they produce on the treble recorder. This means that sopranino players can use Part II exactly as it stands. (The sounds will then be at the correct octave for the sopranino recorder, an octave above the treble recorder.)

Parts for the **descant** recorder are written in the treble clef, but sound an octave higher than the pitch at which they are written. The same fingerings produce notes an octave higher than they produce on the tenor recorder. This means that descant players can use Part I exactly as it stands. (The sounds will then be at the correct octave for the descant recorder, an octave above the tenor recorder.)

Parts for the **bass** recorder are written in the bass clef, and sound an octave higher than the pitch at which they are written. The same fingerings produce notes an octave lower than they produce on the treble recorder. This means that bass players can use Part II as it stands, provided that they are able to read from the treble clef, and to play the notes with the same fingerings as they would use on the treble. (The sounds will then be at the correct octave for the bass recorder, an octave below the treble recorder.) Since players do not ordinarily *begin* with the bass recorder, but learn the smaller instruments first, this should give them no special difficulty; and we have not gone to the expense of printing Part II all over again in the bass clef. (The very highest notes are in practice unobtainable on most bass recorders, and should be left out. This does not matter, since they never occur in bass recorder parts.)

NOTATION OF REGISTERS

The HELMHOLTZ Notation is used throughout this book, as shown in the following example:

This will indicate the written pitch, but not necessarily the sounding pitch, since the sopranino, descant and bass recorders are instruments sounding the octave above written pitch.

SCALES AND ARPEGGIOS
FOR DESCANT AND TENOR RECORDERS

Part of the Scale of G major. *Very Easy* (1)

Part of the Scale of D major. *Very Easy* (2)

Scale of D major. *Very Easy* (4)

Common Chord of D major. *Very Easy* (3)

Scale of C major. *Easy* (7)

Common Chord of C major. *Easy* (5)

Scale of E minor. *Easy* (8)

Note: The figures set by the grading of each scale or broken chord refer to the special exercises beginning on page 26 (see *Using the special exercises* page 2)

SCALES AND ARPEGGIOS FOR DESCANT AND TENOR RECORDERS

Common Chord of E minor. *Easy* (9)

Scale of G major. *Easy* (1, 10)

Common Chord of G major. *Easy* (10)

Scale of F major. *Easy* (6, 11, 12, 13, 14)

Common Chord of F major. *Easy* (14)

Scale of D minor. *Easy* (11, 12, 13, 14, 34, 35)

Common Chord of D minor. *Easy* (7)

Scale of G minor. *Easy* (20, 29, 30)

SCALES AND ARPEGGIOS FOR DESCANT AND TENOR RECORDERS

Common Chord of G minor. *Easy* (16)

Scale of A major. *Easy* (17)

Common Chord of A major. *Easy* (17)

Scale of A minor. *Easy* (6, 11, 12, 13, 15)

Common Chord of A minor. *Easy* (15)

Scale of D major. *Moderate* (4, 22, 23)

Scale of A major. *Moderate* (17, 21, 22, 23)

SCALES AND ARPEGGIOS FOR DESCANT AND TENOR RECORDERS

Scale of D minor (Melodic). *Moderate* (29, 30, 34, 35)

Scale of D minor (Harmonic). *Moderate* (29, 30, 34, 35)

Common Chord of E minor. *Moderate* (31)

Common Chord of E major. *Moderate* (32)

Scale of A minor (Harmonic). *Moderate* (15, 24, 32)

Scale of A minor (Melodic). *Moderate* (17, 22, 23)

Common Chord of A major. *Moderate* (17, 27, 34, 35)

SCALES AND ARPEGGIOS FOR DESCANT AND TENOR RECORDERS

Common Chord of F sharp minor. *Moderate* (27)

Scale of E minor (Harmonic). *Moderate* (18, 19, 28)

Scale of E minor (Melodic). *Moderate* (18, 28)

Dominant Seventh of D. *Moderate* (27, 34, 35)

Common Chord of A minor. *Fairly Difficult* (26, 38, 39)

Scale of G major. *Fairly Difficult* (53)

Common Chord of D major. *Fairly Difficult* (53)

SCALES AND ARPEGGIOS FOR DESCANT AND TENOR RECORDERS

Scale of B minor (Melodic). *Fairly Difficult* (33)

Common Chord of B minor. *Fairly Difficult* (53)

Dominant Seventh of A. *Fairly Difficult* (53)

Scale of F major. *Fairly Difficult* (48)

Common Chord of F major. *Fairly Difficult* (26, 47, 53)

Dominant Seventh of F. *Fairly Difficult* (47, 48)

Common Chord of B flat major. *Fairly Difficult* (48, 49)

SCALES AND ARPEGGIOS FOR DESCANT AND TENOR RECORDERS

Dominant Seventh of C. *Fairly Difficult* (53)

Diminished Seventh of C. *Fairly Difficult* (24, 53)

Scale of E major. *Fairly Difficult* (40, 41, 42)

Common Chord of C sharp minor. *Fairly Difficult* (27, 34, 35)

Diminished Seventh of A. *Fairly Difficult* (24, 53)

Diminished Seventh of D. *Fairly Difficult* (34, 35)

Scale of B flat major. *Fairly Difficult* (43, 44, 45, 48)

SCALES AND ARPEGGIOS FOR DESCANT AND TENOR RECORDERS

Diminished Seventh of F. *Fairly Difficult* (34, 35)

Chromatic Scale. *Fairly Difficult* (12, 20, 28, 33, 35, 39, 45, 47, 51, 52, 53, 54)

Scale of B major. *Difficult* (40, 41, 42, 46)   Enharmonic C flat

Dominant Seventh of F sharp. *Difficult* (50, 51, 55, 56)   Enharmonic G flat

Common Chord of F minor. *Difficult* (44, 47, 60)

Scale of F sharp minor (Harmonic). *Difficult* (50, 51)

SCALES AND ARPEGGIOS FOR DESCANT AND TENOR RECORDERS

Scale of F sharp minor. (Melodic). *Difficult* (50, 51)

Dominant Seventh of B flat. *Difficult* (25, 57, 60)

Common Chord of C minor. *Difficult* (25, 38, 43, 44, 45)

Diminished Seventh of G. *Difficult* (25, 36, 37, 57)

Scale of E flat major. *Difficult* (43, 45, 48)

Dominant Seventh of E flat. *Difficult* (48)

Common Chord of B flat minor *Difficult* (54, 58, 59)

SCALES AND ARPEGGIOS FOR DESCANT AND TENOR RECORDERS

Dominant Seventh of B. *Difficult* (46, 55)

Diminished Seventh of B. *Difficult* (46)

Dominant Seventh of E. *Difficult* (40, 42, 57)

Common Chord of B major. *Difficult* (40, 42, 57)

Diminished Seventh of E. *Difficult* (40, 42, 57)

Scale of G minor (Melodic). *Difficult* (43, 44, 45)

Scale of G minor (Harmonic). *Difficult* (20, 36, 37)

SCALES AND ARPEGGIOS FOR DESCANT AND TENOR RECORDERS

Common Chord of E flat major. *Difficult* (57, 60)

Common Chord of D flat major. *Difficult* (55, 56, 58, 59)

Diminished Seventh of E flat. *Difficult* (53)

Common Chord of F sharp major. *Difficult* (55, 56)

Diminished Seventh of F sharp. *Difficult* (58, 59)

Scale of C sharp minor (Harmonic). *Difficult* (40, 41, 42, 55, 56)

Scale of C sharp minor. (Melodic). *Difficult* (40, 41, 42, 55, 56)

SCALES AND ARPEGGIOS FOR DESCANT AND TENOR RECORDERS

Common Chord of A flat major. *Difficult* (44, 45, 57, 60)

Common Chord of G sharp minor. *Difficult* (46, 56, 57, 60)

Scale of F minor (Harmonic). *Difficult* (48, 54, 58, 59)

Scale of F minor (Melodic). *Difficult* (48, 54, 58, 59)

Diminished Seventh of B flat. *Difficult* (36, 37, 57, 60)

Common Chord of E flat minor. *Difficult* (36, 37, 57, 60)

Scale of A flat major. *Difficult* (58, 59)

SCALES AND ARPEGGIOS FOR DESCANT AND TENOR RECORDERS

Dominant Seventh of A flat. *Difficult* (36, 37, 58, 59)

Diminished Seventh of A flat. *Difficult* (54)

Scale of G sharp minor (Harmonic). *Very Difficult* (46, 55, 56)

Enharmonic
A flat minor

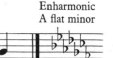

Scale of G sharp minor (Melodic). *Very Difficult* (46, 55, 56)

Enharmonic
A flat minor

Scale of C minor (Melodic). *Very Difficult* (36, 37, 43, 45, 57, 60)

Scale of C minor (Harmonic). *Very Difficult* (36, 37, 43, 45, 57, 60)

Diminished Seventh of D flat. *Very Difficult* (43, 44, 45, 60)

EXERCISES FOR DESCANT AND TENOR RECORDERS
with notes on the special difficulties encountered in them

VERY EASY

1

To avoid an extra note in changing from b' to c'' and from c'' to b' concentrate on lifting the finger which is coming up promptly.

VERY EASY

2

When learning to use the right hand for the first time, make sure that the fingers of the left hand do not move. They should, as usual, be kept relaxed, but firm, with the soft pad covering the holes. It is natural to imagine that the fault is with the fingers of the right hand which you are newly learning to use. But it may be the left-hand fingers which are getting out of place, because you are now concentrating on the right hand.

In changing from e' to f♯' and from f♯' to e', use the second finger of the right hand as a pivot, and allow a slight rotary movement of the wrist.

VERY EASY

3

In changing from a' to d'', keep a firm hold with the second finger of the left hand and release the thumb and first finger simultaneously (supporting the instrument with the right hand). Relax the breath pressure as you go down to the low d'.

VERY EASY

4

In changing from b' to c♯'' and from c♯'' to b', use the first finger of the left hand as a pivot, and allow a slight rotary movement of the wrist.

EXERCISES FOR DESCANT AND TENOR RECORDERS

EASY

5

In playing repeated notes make sure that the tonguing is clean and neat.
    The bottom joint of your recorder should be turned a little to the right so that your little finger can cover its hole without either stretching out or pulling back. Breathe very gently into the instrument to sound the bottom c¹.

EASY

6

In changing from d" to e" the thumb joint should be bent *before* it approaches the recorder, so that the nail rides across the hole. Make sure that the fingers arrive at exactly the same moment as the thumb.
    Both the change from e" to c" and the change from e" to b¹ need the same thumb movement. This movement should be made in contact with the recorder. The thumb should not jump away. The joint should be straightened with a very definite movement so that the pad again covers the hole.
    To avoid an extra note in changing from e" to f" and from f" to e", concentrate on lifting the finger which is coming up.
    To make a clean change from e" to d", feel that the recorder is well supported by the second finger of the left hand and the thumb of the right hand. Let the mouth-piece rest on your lower lip: do not grip it between your teeth. Lift all the other fingers and the left thumb simultaneously.
    To avoid an extra note in changing from d" to b¹ lift the second finger of the left hand promptly, and pinch the thumb and first finger together.

EASY

7

In changing from g¹ to f¹, get the fingers of the right hand in position, with the second finger raised; then put them down simultaneously, concentrating on the little finger.
    To avoid an extra note in changing from f¹ to e¹ concentrate on lifting the third and fourth fingers of the right hand simultaneously.

To avoid an extra note in changing from e' to f' lift the second finger of the right hand promptly, and make the third and fourth fingers go down as one.

In the change from f' to d', and from d' to f', the second finger and the little finger of the right hand alternate in their movements. Notice this movement, and practise it.

EASY

In playing the octaves from e' to e'' and from e'' to e', the thumb only should move. Once again, this movement should be made in contact with the recorder. The thumb should not jump away. On the other hand, it should not merely remain pressed flat against the recorder, and be slid up or down. The joint should be bent (for the upper note) or straightened (for the lower note) with a very definite movement.

EASY

The changes from b' to e'', e'' to b', e'' to c'', and c'' to e'' should be made with a very definite thumb movement. At the same time the necessary fingers should be lifted together.

EASY

In changing from d'' to g'', the thumb joint should be bent before it approaches the recorder.

In playing the octaves from g' to g'' and from g'' to g', the thumb only should move, with a very definite movement of the joint.

EASY

EXERCISES FOR DESCANT AND TENOR RECORDERS

EASY

EASY

These three exercises give practice in changing quickly from b' to c'', d'' to e'', e'' to f'', and e' to f'.

EASY

In changing from c'' to f'' and from f'' to c'' the thumb should move with a very definite bending or straightening of the joint.

In changing from a' to b♭', make sure that the second finger of the left hand does not stick and get left behind.

In changing from b♭' to c'', think of the fingers coming up rather than of the finger going down.

In changing from c'' to b♭', think of lifting the second finger of the left hand. Have the first finger of the right hand poised ready to go down with the first and third fingers of the left hand.

In changing from b♭' to a', think of the two fingers coming up rather than of the finger going down.

EASY

In changing from f'' to g♯'', the third fingers of both hands come up. Nothing else moves.

EXERCISES FOR DESCANT AND TENOR RECORDERS

EASY

16

In changing from g' to b♭', the second finger of the left hand must be lifted promptly, as the first finger of the right hand is put down.

In changing from b♭' to d'', the second finger of the left hand fills the gap, while the other fingers and the thumb come up.

In changing from d'' to b♭', the second finger of the left hand must come up promptly, as the first and third fingers and the thumb of the left hand, and the first finger of the right hand, go down.

EASY

17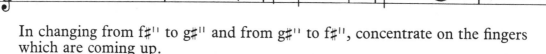

In changing from f♯'' to g♯'' and from g♯'' to f♯'', concentrate on the fingers which are coming up.

High a'' is sometimes a more difficult note to get than the notes immediately above it. Try experiments with the size of the thumb hole, and with the breath pressure and tonguing, until you find the best way to play a sweet, easy note.

EASY

18

EASY

19

In changing from c'' to d♯'', the fingers going down should move simultaneously. Let them seem to be pushing the thumb off.

In changing back, the reverse is the case: the thumb pushes the fingers off. But hold on to the second finger of the left hand.

In changing from b' to d♯'', the thumb and first finger of the left hand come up. Everything below that, except the little finger of the right hand, goes down simultaneously.

In changing back the reverse is the case: the thumb and first finger of the left hand pinch together. Everything else comes up.

In changing from c♯'' to d♯'', the first finger of the left hand comes up. The third finger of the left hand and the first, second and third fingers of the right hand go down.

EASY

Notice the alternative fingerings in the chart for e♭'' and d♯''. Decide which fingering gives the best intonation on your recorder in this passage. Choose that fingering.

In changing from e♭'' to f♯'', and from f♯'' to e♭'', concentrate on the fingers which are coming up.

MODERATE

In changing from g♯' to f♯', and from f♯' to g♯', use the second finger of the right hand as a pivot, and allow a slight rotary movement of the wrist.

MODERATE

MODERATE

The change from a'' to b'' may make it necessary to adjust the thumb. Let it be flexible and ready to adapt itself. The b'' may require slightly firmer tonguing.

MODERATE

In changing from g♯' to f', the third and fourth fingers of the right hand and the third finger of the left hand must go down simultaneously, as the second finger of the right hand comes up. Make sure that the little finger is not late.

In changing from f' to g♯', think of the fingers which are coming up.

MODERATE

The change from c'' to e♭'' is the same for the fingers as the change from c'' to d♯''. It only looks different. Occasionally a particular recorder may require the first finger of the left hand to be added for the e♭'' (as well as the other fingers). Sometimes a particular passage needs it in order to be well in tune. Listen carefully, and notice whether this is so. Add the extra finger if required.

MODERATE

Before trying this exercise make sure that the bottom joint of your recorder is turned to the most comfortable position for you. This will depend on the length of your little finger. Turn the joint so that the bottom hole is in the place where your little finger will land neatly on it of its own accord. If it has to stretch out or pull back it will be late in getting there. If it has to stretch out it will also force the other fingers to strain away from their holes.

In playing f' see that the little finger of the right hand is fully covering the bottom hole. Then only the second finger of the right hand needs to be added to give a good c'.

In a drop from any note with few fingers on, to the bottom c', make sure that the lower fingers arrive at the same moment as the upper fingers, and do not arrive late.

EXERCISES FOR DESCANT AND TENOR RECORDERS

In making a big jump up or down, remember to adjust the breath pressure. Very little breath is needed for the low notes. Much more is needed for the upper notes.

MODERATE

In changing from f♯¹ to c♯¹, try covering the bottom hole completely; then draw back the little finger of the right hand extremely quickly. But do this on the recorder rather than in the air.

MODERATE

In changing from e¹ to d♯¹, try the trick of tipping the third finger of the right hand before putting it down; but make sure that the second finger does not tip as well. When playing the f♯¹, the third finger must be quite flat.

In changing from c♯¹¹ to d♯¹¹, listen carefully to the intonation. Then you can decide whether or not to lift the first finger of the left hand (see chart of fingerings). Choose the fingering which sounds best in tune.

MODERATE

MODERATE

These two exercises give further practice in the awkward cross-fingerings between b♭¹ and a¹, and b♭¹ and c¹¹.

MODERATE

31

In changing from g'' to b'', the third finger of the left hand must come up at exactly the moment when the first and second fingers of the right hand go down. Let the thumb be flexible.

MODERATE

32

In changing from e'' to g#'', only one finger of each hand moves.
   In the change from g#'' to b'', one of those fingers (the second of the right hand) goes down again.

MODERATE

33

In changing from a#' to g#', make sure that the third finger of the left hand comes up promptly as the two second fingers go down.
   In changing from g#' to a#' the two second fingers come up as the third finger of the left hand goes down.

MODERATE

34

MODERATE

35

To make a clean change from b♭' to c♯'', the thumb and the third finger of the left hand, and the first finger of the right hand, must come up promptly. The first finger of the left hand remains in position as an anchorage; it is joined by the second finger of the left hand.

In returning from c♯'' to b♭', it is the second finger of the left hand which must come up promptly.

In changing from d' to c♯', try the trick of covering the bottom hole completely and then quickly sliding your little finger back. Try the same method for the change from e' to c♯'.

MODERATE

36

MODERATE

37

In changing from f♯' to e♭', turn your hand slightly towards the thumb in order to uncover the half hole.

In changing from e♭' to f♯', turn your hand in the opposite direction to re-cover the hole.

FAIRLY DIFFICULT

38

FAIRLY DIFFICULT

39

High c''' is one of the hardest notes to get on the recorder. Let the thumb ride flexibly across its hole. On most recorders a very minute opening is required. Make the tonguing firm, with plenty of breath behind it. Practise this note on your own recorder until you have found the best way to make it speak clearly and sweetly. On a different recorder you might have to humour it a little differently.

EXERCISES FOR DESCANT AND TENOR RECORDERS

FAIRLY DIFFICULT

FAIRLY DIFFICULT

FAIRLY DIFFICULT

In changing from f♯' to d♯', turn your hand slightly towards the thumb as the first finger of the right hand goes down, and the third finger tips or slides away from the half hole.

In changing from d♯' to f♯', the turning will be in the opposite direction.

In changing from g♯' to d♯', the third fingers of each hand move—one to cover its hole completely, the other only half—and here the tipped position of the right-hand third finger seems definitely the more effective method.

FAIRLY DIFFICULT

FAIRLY DIFFICULT

FAIRLY DIFFICULT

The change from d' to e♭' (and from c' to e♭') can be done in two ways. You can *slide* back the third finger of the right hand. Or you can *tip* the finger back to the same extent (taking care not to tip the neighbouring fingers as well). Discover which way suits you (and your recorder) best.

In changing from e♭' to f', you can *slide* the third finger of the right hand over the complete hole, or you can *flatten* it. In either case, you must do so exactly as the little finger lands on the bottom hole and the second finger comes up.

In changing from f' to e♭', the half-holing of the third finger, and the lifting of the little finger of the right hand must be done exactly with the movement of the second finger as it goes down.

The changes from f' to a♭' and from a♭' to f' are the same as the changes from f' to g♯' and from g♯' to f' in exercise 24.

The change from a♭' to e♭' is the same for the fingers as the change from g♯' to d♯'. It only looks different.

FAIRLY DIFFICULT

In changing from c♯'' to a♯' lift the second finger promptly.
Try the tipping action for the d♯' here.

FAIRLY DIFFICULT

Another bottom note exercise. Be sure the bottom joint of your recorder is correctly placed, and that all your fingers go down together when there is a big leap.

Try the tipping action of the finger for the e♭' (d♯'), but not for the d♭' (c♯') (the movement of the little finger is too difficult to isolate; the third finger gets involved).

In changing from g♯' to a♯' make sure that the two second fingers come up together.

EXERCISES FOR DESCANT AND TENOR RECORDERS

To play high notes successfully allow the thumb nail to ride flexibly across the back hole. This allows the amount of opening to be adjusted quickly, and with practice almost automatically. A good flow of breath is required.

In changing from b♭" to c"' make sure that the third finger of the right hand comes up promptly. If you forget to bring this finger up, the c"' will be very flat.

Top d"' is usually an easier note to get than top c"', but it is tricky from b♭". Try fingering the melodies silently before blowing them.

In changing from f♯' to e♯' concentrate on lifting the second finger of the right hand promptly.

In changing from e♯' to f♯' the first and little fingers of the right hand must come up promptly together.

In changing from f♯'' to e♯'' make sure that the second finger of the right hand is not late in coming up.

In returning from e♯'' to f♯'' it is the first and third fingers of the right hand which must be prompt.

A great deal of practice is required to make the half-holing of d♯' and c♯' secure; particularly the latter, as the little finger is naturally weak. Decide which method—the *sliding* or the *tipping*—you prefer and stick to it, though it is not necessary to use the same method for the d♯' as for the c♯'. Whichever movement you use be careful not to allow the neighbouring finger to become affected.

DIFFICULT

The chief difficulty of this exercise is to read it quickly and accurately.

In the change from b♯'' to c♯'' the first finger of the left hand should seem to be pushing the thumb off.

DIFFICULT

Another high note exercise. Good breath control and a flexible thumb are required.

In playing difficult slurs increase the breath pressure slightly on the second note, and make sure that the fingers move at exactly the same time.

In changing from c''' to d''' lift the second finger of the right hand promptly.

In changing from d''' to c''' the two third fingers come up together while the second finger of the right hand goes down.

Notice that in the change from b'' to d''' the movement of the two hands is the same. This also applies to the change from d''' to b''.

EXERCISES FOR DESCANT AND TENOR RECORDERS

VERY DIFFICULT

54

Some more practice with half-holing.

The changes affecting e♭¹ and d♭¹ are the same for the fingers as the changes affecting d♯¹ and c♯¹ in exercises 47 and 48.

In changing from e♭¹¹ to b♭¹ concentrate on the fingers coming up.

VERY DIFFICULT

55

VERY DIFFICULT

56

Here are two more exercises which include half-holing.

In changing from a♯¹ to c♯¹¹ concentrate on the fingers coming up.

In changing from c♯¹ to e♯¹ and back it is not necessary to move the little finger. The e♯¹ will be sufficiently well in tune with the bottom hole only half covered.

The changes from d♯¹ to c♯¹ and back, and from d♯¹ to e♯¹ and back, are particularly difficult, and require careful and constant practice.

EXERCISES FOR DESCANT AND TENOR RECORDERS

VERY DIFFICULT

57

High eb''' is inclined to be shrill, but with careful adjustment of the thumb—usually a rather larger aperture than for d'''—this note should be obtainable.

In changing from bb'' to eb''' lift all the fingers of the right hand and the second finger of the left hand promptly, the third finger of the left hand only going down.

In changing from eb''' to bb'' the first, second and third fingers of the right hand and the second finger of the left hand go down, while the third finger of the left hand comes up.

VERY DIFFICULT

58

VERY DIFFICULT

59

Another two exercises in half-holing.

The difficult changes from db' to eb' and from eb' to db' have already been practised as c#' to d#' and d#' to c#', but they are probably still far from perfect.

In changing from f' to db' slide the little finger of the right hand half back as the second finger comes down.

41

EXERCISES FOR DESCANT AND TENOR RECORDERS

VERY DIFFICULT

This is the hardest exercise in the book. In such big leaps, it is very difficult to adjust the thumb, the breath pressure and the tonguing quickly enough. The chief secret is to keep relaxed. At the same time the fingers have to come down firmly. You will need plenty of power behind your breathing: what singers call "support". It is also a great help to look well ahead so that you are not taken by surprise.

SCALES AND ARPEGGIOS
FOR SOPRANINO, TREBLE, AND BASS RECORDERS

Part of the Scale of C major. *Very Easy* (1)

Part of the Scale of G major. *Very Easy* (2)

Scale of G major. *Very Easy* (4)

Common Chord of G major. *Very Easy* (3)

Scale of F major. *Easy* (7)

Common Chord of F major. *Easy* (5)

Scale of A minor. *Easy* (8)

Note: The figures set by the grading of each scale or broken chord refer to the special exercises beginning on page 58 (see *Using the special exercises* page 2)

SCALES AND ARPEGGIOS FOR SOPRANINO, TREBLE, AND BASS RECORDERS

Common Chord of A minor. *Easy* (9)

Scale of C major. *Easy* (1, 10)

Common Chord of C major. *Easy* (10)

Scale of B flat major. *Easy* (6, 11, 12, 13, 14)

Common Chord of B flat major. *Easy* (14)

Scale of G minor. *Easy* (11, 12, 13, 14, 34, 35)

Common Chord of G minor. *Easy* (7)

Scale of C minor. *Easy* (20, 29, 30)

SCALES AND ARPEGGIOS FOR SOPRANINO, TREBLE, AND BASS RECORDERS

Scale of G minor (Melodic). *Moderate* (29, 30, 34, 35)

Scale of G minor (Harmonic). *Moderate* (29, 30, 34, 35)

Common Chord of A minor. *Moderate* (31)

Common Chord of A major. *Moderate* (32)

Scale of D minor (Harmonic). *Moderate* (15, 24, 32)

Scale of D minor (Melodic). *Moderate* (17, 22, 23)

Common Chord of D major. *Moderate* (17, 27, 34, 35)

SCALES AND ARPEGGIOS FOR SOPRANINO, TREBLE, AND BASS RECORDERS

Common Chord of B minor. *Moderate* (27)

Scale of A minor (Harmonic). *Moderate* (18, 19, 28)

Scale of A minor (Melodic). *Moderate* (18, 28)

Dominant Seventh of G. *Moderate* (27, 34, 35)

Common Chord of D minor. *Fairly Difficult* (26, 38, 39)

Scale of C major. *Fairly Difficult* (53)

Common Chord of G major. *Fairly Difficult* (53)

SCALES AND ARPEGGIOS FOR SOPRANINO, TREBLE, AND BASS RECORDERS

Common Chord of G minor. *Fairly Difficult* (53)

Scale of F major. *Fairly Difficult* (39, 53)

Common Chord of F major. *Fairly Difficult* (38, 39, 53)

Dominant Seventh of C. *Fairly Difficult* (53)

Common Chord of C major. *Fairly Difficult* (53)

Common Chord of C minor. *Fairly Difficult* (49)

Scale of E minor (Harmonic). *Fairly Difficult* (33)

SCALES AND ARPEGGIOS FOR SOPRANINO, TREBLE, AND BASS RECORDERS

Scale of E minor (Melodic). *Fairly Difficult* (33)

Common Chord of E minor. *Fairly Difficult* (53)

Dominant Seventh of D. *Fairly Difficult* (53)

Scale of B flat major. *Fairly Difficult* (48)

Common Chord of B flat major. *Fairly Difficult* (26, 47, 53)

Dominant Seventh of B flat. *Fairly Difficult* (47, 48)

Common Chord of E flat major. *Fairly Difficult* (48, 49)

Dominant Seventh of F. *Fairly Difficult* (53)

Diminished Seventh of F. *Fairly Difficult* (24, 53)

Scale of A major. *Fairly Difficult* (40, 41, 42)

Common Chord of F sharp minor. *Fairly Difficult* (27, 34, 35)

Diminished Seventh of D. *Fairly Difficult* (24, 53)

Diminished Seventh of G. *Fairly Difficult* (34, 35)

Scale of E flat major. *Fairly Difficult* (43, 44, 45, 48)

SCALES AND ARPEGGIOS FOR SOPRANINO, TREBLE, AND BASS RECORDERS

Diminished Seventh of B flat. *Fairly Difficult* (34, 35)

Chromatic Scale. *Fairly Difficult* (12, 20, 28, 33, 35, 39, 45, 47, 51, 52, 53, 54)

Scale of E major. *Difficult* (40, 41, 42, 46)

Dominant Seventh of B. *Difficult* (50, 51, 55, 56)

Common Chord of B flat minor. *Difficult* (44, 47, 60)

Scale of B minor (Harmonic). *Difficult* (50, 51)

SCALES AND ARPEGGIOS FOR SOPRANINO, TREBLE, AND BASS RECORDERS

Scale of B minor (Melodic). *Difficult* (50, 51)

Dominant Seventh of E flat. *Difficult* (25, 57, 60)

Common Chord of F minor. *Difficult* (25, 38, 43, 44, 45)

Diminished Seventh of C. *Difficult* (25, 36, 37, 57)

Scale of A flat major. *Difficult* (43, 45, 48)

Dominant Seventh of A flat. *Difficult* (48)

Common Chord of E flat minor. *Difficult* (54, 58, 59)

SCALES AND ARPEGGIOS FOR SOPRANINO, TREBLE, AND BASS RECORDERS

Dominant Seventh of E. *Difficult* (46, 55)

Diminished Seventh of E. *Difficult* (46)

Dominant Seventh of A. *Difficult* (40, 42, 57)

Common Chord of E major. *Difficult* (40, 42, 57)

Diminished Seventh of A. *Difficult* (40, 42, 57)

Scale of C minor (Melodic). *Difficult* (43, 44, 45)

Scale of C minor (Harmonic). *Difficult* (20, 36, 37)

SCALES AND ARPEGGIOS FOR SOPRANINO, TREBLE, AND BASS RECORDERS

Common Chord of A flat major. *Difficult* (57, 60)

Common Chord of F sharp major. *Difficult* (55, 56, 58, 59)

Diminished Seventh of A flat. *Difficult* (53)

Common Chord of B major. *Difficult* (55, 56)

Diminished Seventh of B. *Difficult* (58, 59)

Scale of F sharp minor (Harmonic). *Difficult* (40, 41, 42, 55, 56)

Scale of F sharp minor (Melodic). *Difficult* (40, 41, 42, 55, 56)

SCALES AND ARPEGGIOS FOR SOPRANINO, TREBLE, AND BASS RECORDERS

Common Chord of D flat major. *Difficult* (44, 45, 57, 60)

Common Chord of C sharp minor. *Difficult* (46, 56, 57, 60)

Scale of B flat minor (Harmonic). *Difficult* (48, 54, 58, 59)

Scale of B flat minor (Melodic). *Difficult* (48, 54, 58, 59)

Diminished Seventh of E flat. *Difficult* (36, 37, 57, 60)

Common Chord of G sharp minor. *Difficult* (36, 37, 57, 60)

Scale of D flat major. *Difficult* (58, 59)

Enharmonic C sharp major

SCALES AND ARPEGGIOS FOR SOPRANINO, TREBLE, AND BASS RECORDERS

Dominant Seventh of D flat. *Difficult* (36, 37, 58, 59)

Diminished Seventh of D flat. *Difficult* (54)

Scale of C sharp minor (Harmonic). *Very Difficult* (46, 55, 56)

Scale of C sharp minor (Melodic). *Very Difficult* (46, 55, 56)

Scale of F minor (Melodic). *Very Difficult* (36, 37, 43, 45, 57, 60)

Scale of F minor (Harmonic). *Very Difficult* (36, 37, 43, 45, 57, 60)

Diminished Seventh of F sharp. *Very Difficult* (43, 44, 45, 60)

EXERCISES FOR
SOPRANINO, TREBLE, AND BASS RECORDERS
with notes on the special difficulties encountered in them

VERY EASY

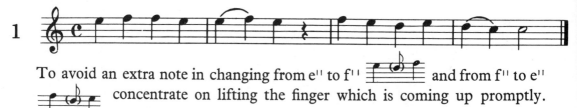

To avoid an extra note in changing from e'' to f'' and from f'' to e'' concentrate on lifting the finger which is coming up promptly.

VERY EASY

When learning to use the right hand for the first time, make sure that the fingers of the left hand do not move. They should, as usual, be kept relaxed, but firm, with the soft pad covering the holes. It is natural to imagine that the fault is with the fingers of the right hand which you are newly learning to use. But it may be the left-hand fingers which are getting out of place, because you are now concentrating on the right hand.

In changing from a' to b' and from b' to a', use the second finger of the right hand as a pivot, and allow a slight rotary movement of the wrist.

VERY EASY

In changing from d'' to g'' keep a firm hold with the second finger of the left hand and release the thumb and first finger simultaneously (supporting the instrument with the right hand). Relax the breath pressure as you go down to the low g'.

VERY EASY

In changing from e'' to f♯'', and from f♯'' to e'' use the first finger of the left hand as a pivot, and allow a slight rotary movement of the wrist.

EXERCISES FOR SOPRANINO, TREBLE, AND BASS RECORDERS

EASY

5

In playing repeated notes make sure that the tonguing is clean and neat.

The bottom joint of your recorder should be turned a little to the right so that your little finger can cover its hole without either stretching out or pulling back. Breathe very gently into the instrument to sound the bottom f'.

EASY

6

In changing from g'' to a'' the thumb joint should be bent *before* it approaches the recorder so that the nail rides across the hole. Make sure that the fingers arrive at exactly the same moment as the thumb.

Both the change from a'' to f'' and the change from a'' to e'' need the same thumb movement. This movement should be made in contact with the recorder. The thumb should not jump away. The joint should be straightened with a very definite movement so that the pad again covers the hole.

To avoid an extra note in changing from a'' to b♭'' and from b♭'' to a'', concentrate on lifting the finger which is coming up.

To make a clean change from a'' to g'', feel that the recorder is well supported by the second finger of the left hand and the thumb of the right hand. Let the mouth-piece rest on your lower lip: do not grip it between your teeth. Lift all the other fingers and the left thumb simultaneously.

To avoid an extra note in changing from g'' to e'' lift the second finger of the left hand promptly, and pinch the thumb and first finger together.

EASY

7

In changing from c'' to b♭', get the fingers of the right hand in position, with the second finger raised; then put them down simultaneously, concentrating on the little finger.

To avoid an extra note in changing from b♭' to a' concentrate on lifting the third and fourth fingers of the right hand simultaneously.

To avoid an extra note in changing from a' to b♭'  lift the second finger of the right hand promptly, and make the third and fourth fingers go down as one.

In the change from b♭' to g', and from g' to b♭', the second finger and the little finger of the right hand alternate in their movements. Notice this movement, and practise it.

EASY

In playing the octaves from a' to a'' and from a'' to a', the thumb only should move. Once again, this movement should be made in contact with the recorder. The thumb should not jump away. On the other hand, it should not merely remain pressed flat against the recorder, and be slid up or down. The joint should be bent (for the upper note) or straightened (for the lower note) with a very definite movement.

EASY

The changes from e'' to a'', a'' to e'', a'' to f'', and f'' to a'' should be made with a very definite thumb movement. At the same time the necessary fingers should be lifted together.

EASY

In changing from g'' to c''' the thumb joint should be bent before it approaches the recorder.

In playing the octaves from c'' to c''' and from c''' to c'', the thumb only should move, with a very definite movement of the joint.

EASY

EXERCISES FOR SOPRANINO, TREBLE, AND BASS RECORDERS

EASY

12

EASY

13

These three exercises give practice in changing quickly from e'' to f'', g'' to a'', a'' to b♭'', and a' to b♭'.

EASY

14

In changing from f'' to b♭'' and from b♭'' to f'', the thumb should move with a very definite bending or straightening of the joint.

In changing from d'' to e♭'', make sure that the second finger of the left hand does not stick and get left behind.

In changing from e♭'' to f'', think of the fingers coming up rather than of the finger going down.

In changing from f'' to e♭'', think of lifting the second finger of the left hand. Have the first finger of the right hand poised ready to go down with the first and third fingers of the left hand.

In changing from e♭'' to d'', think of the two fingers coming up rather than of the finger going down.

EASY

15

In changing from b♭'' to c♯''', the third fingers of both hands come up. Nothing else moves.

EXERCISES FOR SOPRANINO, TREBLE, AND BASS RECORDERS

EASY

In changing from c'' to e♭'', the second finger of the left hand must be lifted promptly, as the first finger of the right hand is put down.

In changing from e♭'' to g'', the second finger of the left hand fills the gap, while the other fingers and the thumb come up.

In changing from g'' to e♭'', the second finger of the left hand must come up promptly, as the first and third fingers and the thumb of the left hand, and the first finger of the right hand, go down.

EASY

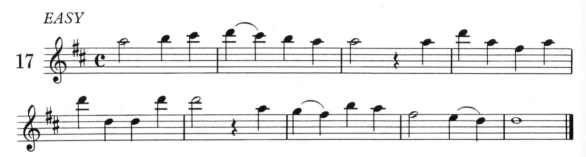

In changing from b'' to c♯''' and from c♯''' to b'', concentrate on the fingers which are coming up.

High d''' is sometimes a more difficult note to get than the notes immediately above it. Try experiments with the size of the thumb hole, and with the breath pressure and tonguing, until you find the best way to play a sweet, easy note.

EASY

EASY

In changing from f'' to g♯'', the fingers going down should move simultaneously. Let them seem to be pushing the thumb off.

In changing back, the reverse is the case: the thumb pushes the fingers off. But hold on to the second finger of the left hand.

In changing from e'' to g♯'', the thumb and first finger of the left hand come up. Everything below that, except the little finger of the right hand, goes down simultaneously.

In changing back the reverse is the case: the thumb and first finger of the left hand pinch together. Everything else comes up.

EXERCISES FOR SOPRANINO, TREBLE, AND BASS RECORDERS

In changing from f♯" to g♯", the first finger of the left hand comes up. The third finger of the left hand and the first, second and third fingers of the right hand go down.

EASY

Notice the alternative fingerings in the chart for a♭" and g♯". Decide which fingering gives the best intonation on your recorder in this passage. Choose that fingering.

In changing from a♭" to b", and from b" to a♭", concentrate on the fingers which are coming up.

MODERATE

In changing from c♯" to b', and from b' to c♯", use the second finger of the right hand as a pivot, and allow a slight rotary movement of the wrist.

MODERATE

MODERATE

The change from d''' to e''' may make it necessary to adjust the thumb. Let it be flexible and ready to adapt itself. The e''' may require slightly firmer tonguing.

EXERCISES FOR SOPRANINO, TREBLE, AND BASS RECORDERS

MODERATE

In changing from c♯" to b♭', the third and fourth fingers of the right hand and the third finger of the left hand must go down simultaneously, as the second finger of the right hand comes up. Make sure that the little finger is not late.

In changing from b♭' to c♯", think of the fingers which are coming up.

MODERATE

The change from f" to a♭" is the same for the fingers as the change from f" to g♯". It only looks different. Occasionally a particular recorder may require the first finger of the left hand to be added for the a♭" (as well as the other fingers). Sometimes a particular passage needs it in order to be well in tune. Listen carefully, and notice whether this is so. Add the extra finger if required.

MODERATE

Before trying this exercise make sure that the bottom joint of your recorder is turned to the most comfortable position for you. This will depend on the length of your little finger. Turn the joint so that the bottom hole is in the place where your little finger will land neatly on it of its own accord. If it has to stretch out or pull back it will be late in getting there. If it has to stretch out it will also force the other fingers to strain away from their holes.

In playing b♭' see that the little finger of the right hand is fully covering the bottom hole. Then only the second finger of the right hand needs to be added to give a good f'.

In a drop from any note with few fingers on, to the bottom f', make sure that the lower fingers arrive at the same moment as the upper fingers, and do not arrive late.

In making a big jump up or down, remember to adjust the breath pressure. Very little breath is needed for the low notes. Much more is needed for the upper notes.

MODERATE

In changing from b¹ to f♯¹, try covering the bottom hole completely; then draw back the little finger of the right hand extremely quickly. But do this on the recorder rather than in the air.

MODERATE

In changing from a¹ to g♯¹, try the trick of tipping the third finger of the right hand before putting it down, but make sure that the second finger does not tip as well. When playing the b¹, the third finger must be quite flat.

In changing from f♯'' to g♯'', listen carefully to the intonation. Then you can decide whether or not to lift the first finger of the left hand (see chart of fingerings). Choose the fingering which sounds best in tune.

MODERATE

MODERATE

These two exercises give further practice in the awkward cross-fingerings between e♭'' and d'', and e♭'' and f''

MODERATE

31

In changing from c''' to e''', the third finger of the left hand must come up at exactly the moment when the first and the second fingers of the right hand go down Let the thumb be flexible.

MODERATE

32

In changing from a'' to c♯''', only one finger of each hand moves.
  In the change from c♯''' to e''' one of those fingers (the second of the right hand) goes down again.

MODERATE

33

In changing from d♯'' to c♯'', make sure that the third finger of the left hand comes up promptly as the two second fingers go down.
  In changing from c♯'' to d♯'' the two second fingers come up as the third finger of the left hand goes down.

MODERATE

34

MODERATE

35

To make a clean change from e♭" to f♯", the thumb and the third finger of the left hand, and the first finger of the right hand, must come up promptly. The first finger of the left hand remains in position as an anchorage; it is joined by the second finger of the left hand.

In returning from f♯" to e♭", it is the second finger of the left hand which must come up promptly.

In changing from g' to f♯', try the trick of covering the bottom hole completely and then quickly sliding your little finger back. Try the same method for the change from a' to f♯'.

MODERATE

MODERATE

In changing from b' to a♭', turn your hand slightly towards the thumb in order to uncover the half hole.

In changing from a♭' to b', turn your hand in the opposite direction to re-cover the hole.

FAIRLY DIFFICULT

High f''' is one of the hardest notes to get on the recorder. Let the thumb ride flexibly across its hole. On most recorders a very minute opening is required. Make the tonguing firm, with plenty of breath behind it. Practise this note on your own recorder until you have found the best way to make it speak clearly and sweetly. On a different recorder you might have to humour it a little differently.

In changing from b¹ to g♯¹, turn your hand slightly towards the thumb as the first finger of the right hand goes down, and the third finger tips or slides away from the half hole.

In changing from g♯¹ to b¹, the turning will be in the opposite direction.

In changing from c♯¹¹ to g♯¹, the third fingers of each hand move—one to cover its hole completely, the other only half—and here the tipped position of the right-hand third finger seems definitely the more effective method.

EXERCISES FOR SOPRANINO, TREBLE, AND BASS RECORDERS

The change from g' to a♭' (and from f' to a♭') can be done in two ways. You can *slide* back the third finger of the right hand. Or you can *tip* the finger back to the same extent (taking care not to tip the neighbouring fingers as well). Discover which way suits you (and your recorder) best.

In changing from a♭' to b♭', you can *slide* the third finger of the right hand over the complete hole, or you can *flatten* it. In either case, you must do so exactly as the little finger lands on the bottom hole and the second finger comes up.

In changing from b♭' to a♭', the half-holing of the third finger, and the lifting of the little finger of the right hand must be done exactly with the movement of the second finger as it goes down.

The changes from b♭' to d♭'' and from d♭'' to b♭' are the same as the changes from b♭' to c♯'' and from c♯'' to b♭' in exercise 24.

The change from d♭'' to a♭', is the same for the fingers as the change from c♯'' to g♯' It only looks different.

FAIRLY DIFFICULT

46

In changing from f♯'' to d♯'' lift the second finger promptly.
Try the tipping action for the g♯' here.

FAIRLY DIFFICULT

47

Another bottom note exercise. Be sure the bottom joint of your recorder is correctly placed, and that all your fingers go down together when there is a big leap.

Try the tipping action of the finger for the a♭' (g♯'), but not for the g♭' (f♯') (the movement of the little finger is too difficult to isolate; the third finger gets involved).

In changing from c♯'' to d♯'' make sure that the two second fingers come up together.

EXERCISES FOR SOPRANINO, TREBLE, AND BASS RECORDERS

DIFFICULT

DIFFICULT

To play high notes successfully allow the thumb nail to ride flexibly across the back hole. This allows the amount of opening to be adjusted quickly, and with practice almost automatically. A good flow of breath is required.

In changing from e♭''' to f''', make sure that the third finger of the right hand comes up promptly. If you forget to bring this finger up, the f''' will be very flat.

Top g''' is usually an easier note to get than top f''', but it is tricky from e♭'''. Try fingering the melodies silently before blowing them.

DIFFICULT

DIFFICULT

In changing from b' to a♯' concentrate on lifting the second finger of the right hand promptly.

In changing from a♯' to b', the first and little fingers of the right hand must come up promptly together.

In changing from b'' to a♯'', make sure that the second finger of the right hand is not late in coming up.

In returning from a♯'' to b'' it is the first and third fingers of the right hand which must be prompt.

A great deal of practice is required to make the half-holing of g♯' and f♯' secure; particularly the latter, as the little finger is naturally weak. Decide which method—the *sliding* or the *tipping*—you prefer and stick to it, though it is not necessary to use the same method for the g♯' as for the f♯'. Whichever movement you use be careful not to allow the neighbouring finger to become affected.

DIFFICULT

52

The chief difficulty of this exercise is to read it quickly and accurately.

In the change from e♯''' to f♯''', the first finger of the left hand should seem to be pushing the thumb off.

DIFFICULT

53

Another high note exercise. Good breath control and a flexible thumb are required.

In playing difficult slurs, increase the breath pressure slightly on the second note, and make sure that the fingers move at exactly the same time.

In changing from f''' to g''', lift the second finger of the right hand promptly.

In changing from g''' to f''', the two third fingers come up together while the second finger of the right hand goes down.

Notice that in the change from e''' to g''' the movement of the two hands is the same. This also applies to the change from g''' to e'''.

EXERCISES FOR SOPRANINO, TREBLE, AND BASS RECORDERS

VERY DIFFICULT

Some more practice with half-holing.

The changes affecting a♭' and g♭' are the same for the fingers as the changes affecting g♯' and f♯' in exercises 47 and 48.

In changing from a♭'' to e♭'' concentrate on the fingers coming up.

VERY DIFFICULT

VERY DIFFICULT

Here are two more exercises which include half-holing.

In changing from d♯'' to f♯'' concentrate on the fingers coming up.

In changing from f♯' to a♯' and back it is not necessary to move the little finger. The a♯' will be sufficiently well in tune with the bottom hole only half covered.

The changes from g♯' to f♯' and back, and from g♯' to a♯' and back, are particularly difficult, and require careful and constant practice.

EXERCISES FOR SOPRANINO, TREBLE, AND BASS RECORDERS

VERY DIFFICULT

57

High a♭''' is inclined to be shrill, but with careful adjustment of the thumb—usually a rather larger aperture than for g'''—this note should be obtainable.

In changing from e♭''' to a♭''', lift all the fingers of the right hand and the second finger of the left hand promptly, the third finger of the left hand only going down.

In changing from a♭''' to e♭''', the first, second and third fingers of the right hand and the second finger of the left hand go down, while the third finger of the left hand comes up.

VERY DIFFICULT

58

VERY DIFFICULT

59

Another two exercises in half-holing.

The difficult changes from g♭' to a♭' and from a♭' to g♭' have already been practised as f♯' to g♯' and g♯' to f♯', but they are probably still far from perfect.

In changing from b♭' to g♭' slide the little finger of the right hand half back as the second finger comes down.

73

EXERCISES FOR SOPRANINO, TREBLE, AND BASS RECORDERS

VERY DIFFICULT

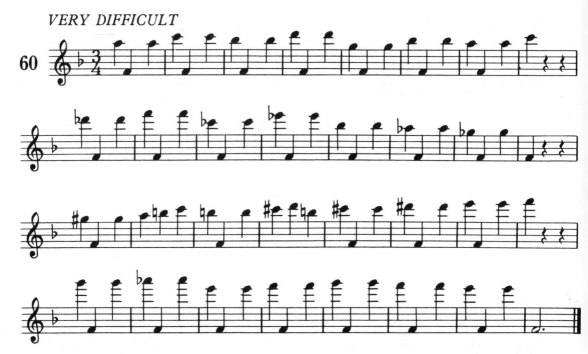

This is the hardest exercise in the book. In such big leaps, it is very difficult to adjust the thumb, the breath pressure and the tonguing quickly enough. The chief secret is to keep relaxed. At the same time the fingers have to come down firmly. You will need plenty of power behind your breathing: what singers call "support". It is also a great help to look well ahead so that you are not taken by surprise.

INDEX

ALTERNATIVE FINGERINGS – 8, 31, 32, 33, 63, 64, 65
ARPEGGIOS (see Broken chords)
ARTICULATION – 6
BASS – 4, 5, 8, 9, 10, 43–74
BOTTOM NOTES – 2, 4, 27, 32, 33, 35, 37, 39, 40, 41, 59, 64, 65, 67, 69, 71, 72, 73
BREATHING – 1, 2, 6, 7, 26, 30, 33, 35, 38, 39, 42, 58, 62, 65, 67, 70, 71, 74
BROKEN CHORDS – 1, 2, 6, 11–25, 43–57
    Common chords of:—
        A flat major – 23, 54
        A major – 13, 14, 46
        A minor – 13, 15, 44, 46
        B flat major – 17, 44, 49
        B flat minor – 20, 51
        B major – 21, 54
        B minor – 17, 47
        C major – 11, 16, 44, 48
        C minor – 20, 45, 48
        C sharp minor – 18, 55
        D flat major – 22, 55
        D major – 11, 15, 45, 46
        D minor – 12, 16, 45, 47
        E flat major – 22, 49
        E flat minor – 23, 52
        E major – 14, 53
        E minor – 12, 14, 49
        F major – 12, 17, 43, 48
        F minor – 19, 52
        F sharp major – 22, 54
        F sharp minor – 15, 50
        G major – 12, 16, 43, 47
        G minor – 13, 16, 44, 48
        G sharp minor – 23, 55
    Diminished sevenths of:—
        A flat – 24, 54
        A – 18, 53
        B flat – 23, 51
        B – 21, 54
        C – 18, 52
        D flat – 24, 56
        D – 18, 50
        E flat – 22, 55
        E – 21, 53
        F – 19, 50
        F sharp – 22, 56
        G – 20, 50
    Dominant sevenths of:—
        A flat – 24, 52
        A – 17, 53
        B flat – 20, 49
        B – 21, 51
        C – 18, 48
        D flat – 25, 56

D – 15, 49
E flat – 20, 52
E – 21, 53
F – 17, 50
F sharp – 19, 57
G flat – See F sharp
G – 16, 47
CHART OF FINGERINGS – 8
CLEFS – 9
CROSS-FINGERINGS – 2, 3, 8, 33, 65
DESCANT – 4, 5, 8, 9, 10, 11–42
DIAPHRAGM CONTROL – 6
DOUBLE-TONGUING – 6
ENHARMONIC NOTES – 2, 19, 24, 25, 32, 37, 40, 41, 55, 57, 64, 69, 72, 73
EXERCISES – 1, 2, 3, 6, 26–42, 58–74
FINGERS – 1, 2, 3, 4, 5, 6, 8, 9, 26–42, 58–74
HALF-HOLING (fingers) – 2, 4, 33, 35, 36, 37, 39, 40, 41, 65, 67, 68, 69, 71, 72, 73
HALF-STOPPING (thumb) – 5, 27, 28, 29, 30, 31, 35, 38, 39, 41, 42, 59, 60, 61, 62, 63, 67, 70, 71, 73, 74
HELMHOLTZ NOTATION – 10
INTONATION – 5, 6, 31, 32, 33, 38, 40, 63, 64, 65, 70, 72
KEYS (on recorders) – 4
KEYS (tonal) – 1, 2
LEAKS – 4
LEFT HAND – 3, 4, 5, 8, 26–42, 58–74
LEGATO – 6
LOUDNESS – 6
MUSCLES – 3, 5
PAD OF THE FINGERS – 4, 5, 26, 27, 58, 59
PHRASING – 6, 7
PITCH – 9, 10
PIVOTING – 4, 5, 26, 31, 58, 63
PRACTICE BOOK FOR THE RECORDER – 1
RECORDER TECHNIQUE – 1
RELAXATION – 3, 5, 6, 26, 42, 58, 74
RIGHT HAND – 3, 4, 5, 8, 26–42, 58–74
ROWLAND-JONES – 1
SCALES – 1, 2, 6, 11–25, 43–57
    A flat major – 23, 52
    A flat minor – See G sharp minor
    A major – 13, 50
    A minor – 13, 14, 43, 47
    B flat major – 18, 44, 49
    B flat minor – 25, 55
    B major – 19, 57
    B minor – 16, 17, 51, 52
    C flat major – See B major
    C major – 11, 16, 43, 44, 47
    C minor – 24, 44, 53
    C sharp major – See D flat major
    C sharp minor – 22, 56

[continued overleaf

INDEX

SCALES *(continued)*
   D flat major – 25, 55
   D major – 11, 13, 45
   D minor – 12, 14, 45, 46
   D sharp minor – See E flat minor
   E flat major – 20, 50
   E flat minor – 25, 57
   E major – 18, 51
   E minor – 11, 15, 48, 49
   F major – 12, 17, 43, 48
   F minor – 23, 56
   F sharp major – 25, 57
   F sharp minor – 19, 20, 54
   G flat major – See F sharp major
   G major – 11, 12, 15, 43, 45
   G minor – 12, 21, 44, 46
   G sharp minor – 24, 57
   Chromatic – 19, 51
SEMITONES – 4

SLIDING – 4, 28, 33, 35, 36, 37, 39, 41, 60, 65, 67, 68, 69, 71, 73
SLURS – 6, 39, 71
SOFTNESS – 6
SOPRANINO – 4, 5, 8, 9, 10, 43–74
STACCATO – 6
STIFFNESS – 3, 5
TENOR – 4, 5, 8, 9, 11–42
THUMB – 4, 5, 8, 26–42, 58–74
TIPPING – 4, 33, 36, 37, 39, 65, 68, 69, 71
TONE – 7
TONGUE – 2, 3, 5, 6, 7, 27, 30, 31, 35, 42, 59, 62, 63, 67, 74
TOP NOTES – 2, 8, 9, 30, 31, 34, 35, 38, 39, 41, 42, 62, 63, 66, 67, 70, 71, 73, 74
TREBLE – 4, 5, 8, 9, 43–74
TRINITY COLLEGE OF MUSIC – 1
TRIPLE-TONGUING – 6